# The Science of How Things

How Do Airplanes Fly?
What Are Black Holes?
Why is the Sky Blue?
How are Clouds Formed?
Planet Earth
What Causes an Earthquake?
Who Discovered Electricity?
How Do Fireflies Light Up?
Why Can Bears Hibernate but Not Humans?
How Do Leaves Change Color?
What Causes Thunder And Lightning?
How Do Magnets Work?
The Moon
How Are Mountains Formed?
What Is Photosynthesis?
How is a Rainbow Formed?
What Are Rain Clouds?
What Causes Seasons?
How Shadows Are Formed?
What Causes Snow?
How Stars Are Formed?
The Sun
What Causes Tornadoes?
How Are Tsunamis Caused?
How Do Volcanoes Form?
What is the Water Cycle?
What Causes Ocean Waves?

# How Do Airplanes Fly?

Airplanes fly thanks to the delicate balance of four forces: lift, thrust, drag, and gravity. The magic starts with the wings, which are specially shaped to create lift. When the plane moves forward, air flows faster over the curved top of the wing than underneath it. This difference in speed creates lower pressure above the wing and higher pressure below, generating lift. This upward force counters gravity, which constantly pulls the plane toward the ground, enabling the airplane to stay aloft.

Thrust, provided by the airplane's engines or propellers, propels the plane forward, overcoming drag, the resistance caused by air pushing back against the moving plane. The faster the airplane moves, the more lift the wings generate. Pilots adjust the angle of the wings, called the angle of attack, to maximize lift without causing too much drag. The intricate interplay of these forces allows an airplane to climb, cruise, and descend, transforming the laws of physics into smooth, controlled flight.

Airplanes embody the principles of aerodynamics, showing how human ingenuity can harness natural forces to achieve incredible feats. From the first powered flights of the Wright brothers to modern jetliners crossing oceans, airplanes demonstrate how understanding science enables us to conquer challenges that once seemed impossible. Every takeoff and landing is a testament to the harmony between engineering and the invisible forces that govern our world.

# What Are Black Holes?

Black holes are some of the most mysterious and powerful objects in the universe, places where gravity is so strong that nothing—not even light—can escape their grasp. They form when massive stars run out of fuel and collapse under their own weight, compressing all their mass into an incredibly small space called a singularity. Surrounding this singularity is the event horizon, the boundary beyond which anything that crosses is lost forever. Once inside, not even the fastest thing in the universe, light, can escape, which is why black holes appear invisible, like a shadow against the backdrop of space.

Despite their invisibility, black holes reveal their presence through their dramatic effects on nearby matter. As they pull in gas, dust, and even stars, this material forms a swirling disk around the black hole, heating up and glowing brightly before disappearing into the void. Black holes can also warp space and time around them, creating phenomena like gravitational lensing, where they bend light from distant stars, and time dilation, where time slows down near the event horizon. These cosmic distortions offer scientists a window into the extreme physics of the universe.

Black holes are not just cosmic destroyers; they are key players in shaping galaxies and the universe itself. Supermassive black holes, millions to billions of times the mass of the Sun, sit at the centers of galaxies and influence their growth and structure. By studying black holes, we learn about the limits of physics and the nature of space, time, and gravity. They are reminders that even in the vastness of the universe, there are places where the laws of nature push reality to its limits, challenging our understanding of how everything works.

# Why is the Sky Blue?

The sky appears blue because of a phenomenon called Rayleigh scattering, which explains how sunlight interacts with Earth's atmosphere. Sunlight, or white light, is made up of all the colors of the spectrum, each with a different wavelength. As sunlight passes through the atmosphere, it encounters molecules and tiny particles that scatter the light in all directions. Shorter wavelengths, like blue and violet, scatter more than longer wavelengths, such as red and yellow. While violet light is scattered even more than blue, our eyes are more sensitive to blue light, and some of the violet is absorbed by the atmosphere, leaving the sky predominantly blue.

The intensity of the blue sky depends on the angle of the Sun and the amount of atmosphere the sunlight travels through. During sunrise and sunset, the Sun is lower in the sky, and its light passes through a thicker layer of atmosphere. This scatters the shorter wavelengths more extensively and allows longer wavelengths, like red and orange, to dominate, creating the warm hues of dawn and dusk. At noon, when the Sun is directly overhead, the shorter blue wavelengths scatter widely, giving the sky its bright, vibrant color.

The blue sky is not just a visual phenomenon; it is a window into the physics of light and the composition of Earth's atmosphere. It demonstrates how even the invisible properties of the air we breathe can shape our experience of the world. The next time you gaze up at a clear blue sky, you are witnessing the delicate interplay of sunlight and air molecules, a reminder of the elegance and complexity woven into the everyday wonders of nature.

# How are Clouds Formed?

Clouds are born from the delicate dance of water, air, and energy in Earth's atmosphere. The process begins when the Sun heats the surface of the Earth, causing water from oceans, lakes, and rivers to evaporate into invisible vapor. This warm, moist air rises into the sky, where it encounters cooler temperatures. As the air cools, the water vapor condenses around tiny particles like dust or salt in the atmosphere, forming millions of tiny water droplets or ice crystals. Together, these droplets or crystals create what we see as a cloud.

The type of cloud that forms depends on how the air rises and cools. If the air gently lifts and spreads out, it creates flat, layered clouds called stratus clouds. If the air rises quickly, often due to strong heat from the ground, it forms towering, fluffy clouds called cumulus clouds. In colder, higher altitudes, clouds might be thin and wispy, known as cirrus clouds, made entirely of ice crystals. Each type of cloud tells a story about the atmosphere's behavior, from calm, steady conditions to powerful updrafts.

Clouds are more than just sky decorations—they are vital players in Earth's water cycle and climate system. They transport water across the planet, provide shade to cool the surface, and reflect sunlight back into space. When clouds become heavy with water droplets, they release precipitation as rain, snow, or hail, returning water to the ground and completing the cycle. Every cloud you see is a dynamic expression of nature's forces, revealing the intricate balance between energy, air, and water that sustains life on Earth.

# Planet Earth

Planet Earth, our home in the cosmos, is a marvel of complexity and beauty, unlike anything we know elsewhere in the universe. Formed about 4.5 billion years ago, Earth emerged from a swirling disk of gas and dust, coalescing into a rocky sphere orbiting the Sun. Its position in the solar system, not too close and not too far from the Sun, places it in the "Goldilocks zone," where temperatures are just right for liquid water to exist—a key ingredient for life. This balance, combined with Earth's protective atmosphere, sets the stage for an extraordinary diversity of environments and ecosystems.

Earth is a dynamic planet, constantly reshaped by internal and external forces. Its molten core drives plate tectonics, causing continents to drift, mountains to rise, and oceans to expand. Meanwhile, its atmosphere and oceans regulate temperatures, distribute energy, and sustain life through the water and carbon cycles. From vast deserts to lush rainforests, icy poles to warm tropics, Earth's surface showcases an intricate interplay of geology, weather, and biology, creating a planet teeming with life. This dynamism makes Earth unique among the rocky worlds of our solar system.

What truly sets Earth apart is its ability to support an extraordinary variety of life forms, from microscopic bacteria to towering trees and intelligent beings—us. Life on Earth thrives through a delicate balance maintained by interconnected systems, where even the smallest changes ripple across ecosystems. Our planet is not just a place we live; it is a living, breathing entity that evolves over time. Understanding Earth means appreciating its fragility and resilience, inspiring us to protect it not just as a home, but as a singular, extraordinary creation in the vastness of the universe.

# What Causes an Earthquake?

Earthquakes are the result of powerful forces deep beneath the Earth's surface, where giant pieces of rock called tectonic plates are constantly moving. These plates make up Earth's crust and sit on a softer layer called the mantle, which allows them to slowly drift over time. Sometimes, these plates get stuck at their edges due to friction, building up enormous pressure as they continue to push against each other. When the pressure becomes too great, it is suddenly released, causing the ground to shake. This release of energy travels through the Earth as seismic waves, which is what we feel as an earthquake.

Most earthquakes happen along the boundaries of tectonic plates, where the plates collide, slide past each other, or pull apart. For example, at a transform boundary like California's San Andreas Fault, plates slide horizontally past one another, creating frequent but often less powerful quakes. In contrast, at convergent boundaries, where plates push together, the collisions can produce massive earthquakes, such as those along the Ring of Fire around the Pacific Ocean. The size of an earthquake depends on how much energy is released, measured by the Richter scale or moment magnitude scale.

Earthquakes remind us of the restless energy within our planet. They shape landscapes by creating mountains, valleys, and even new landforms, but they can also bring destruction when they strike near populated areas. Studying earthquakes helps scientists understand the dynamics of tectonic plates and develop ways to predict and prepare for these natural events. Each earthquake is a reminder of Earth's incredible power and the constant motion beneath our feet, connecting us to the planet's deep and ancient processes.

# Who Discovered Electricity?

Electricity was not discovered by a single person, but rather understood over time through the work of many great minds. Ancient Greeks observed static electricity as early as 600 BCE when rubbing amber, but it wasn't until the 18th and 19th centuries that scientists like Benjamin Franklin, Alessandro Volta, and Michael Faraday made groundbreaking advancements. Their experiments and theories revealed electricity as a fundamental force of nature, laying the foundation for its modern use.

Benjamin Franklin's famous kite experiment was not the discovery of electricity itself, but rather a daring step in humanity's quest to understand it. In the 18th century Benjamin Franklin famously flew a kite into a stormy sky proving that lightning was not the wrath of gods but rather a natural manifestation of electricity.

Electricity is the cosmic dance of charged particles an invisible force that powers our lives and fuels the modern age. It flows through wires like a river of energy carrying the potential to illuminate cities to spark innovation and to drive the engines of industry. At its core electricity is the movement of electrons tiny subatomic particles that orbit the nuclei of atoms. When these electrons move they create an electric current a phenomenon both elegant and profound in its simplicity.

Electricity is a force of nature a manifestation of the fundamental laws of physics that govern our universe. It crackles in the sky during thunderstorms a dazzling reminder of the energy that surrounds us. It courses through our bodies as bioelectric signals that allow us to think to feel to move. Electricity is not merely a tool or a utility it is the language of the cosmos translated into a form we can harness.

Through human ingenuity we have tamed this wild energy and transformed it into a cornerstone of civilization. From Faraday's experiments with electromagnetic induction to Tesla's visionary dreams of alternating current the story of electricity is one of discovery curiosity and boundless potential. It is a story of humanity's quest to understand and to wield the forces that shape the universe.

# How Do Fireflies Light Up?

Fireflies light up through the fascinating process of bioluminescence, a natural chemical reaction that allows them to produce their own light. Inside their tiny abdomens, fireflies have special cells that contain a chemical called luciferin and an enzyme called luciferase. When oxygen combines with luciferin in the presence of luciferase, it creates a reaction that releases energy in the form of light. What makes this process truly remarkable is that it is incredibly efficient, producing almost no heat—a glowing example of nature's ingenuity.

But why do fireflies light up? The answer lies in communication. Fireflies use their flashes to send signals to one another, especially during mating season. Each species of firefly has its own unique pattern of flashes, like a secret code in the dark. Males fly around, blinking their light to attract females, who respond with their own timed flashes. In addition to love signals, some fireflies use their light as a warning, telling predators that they taste bad or are toxic. This dual-purpose glow showcases the evolutionary brilliance of these tiny creatures.

Fireflies remind us of the wonders of nature's design, where even the smallest organisms harness complex chemistry to survive and thrive. Their bioluminescent displays inspire awe and curiosity, teaching us about the connections between biology, chemistry, and physics. The next time you see a firefly's gentle glow, remember that it is not just a pretty light—it is a carefully crafted system, a tiny beacon of life's endless creativity.

# How Was Gravity Discovered?

Gravity, the invisible hand guiding the cosmos, first revealed its secrets through a series of remarkable insights rather than a single epiphany. The story begins with Isaac Newton, the 17th-century physicist and mathematician who saw the universe as a grand machine governed by natural laws. According to legend, Newton was inspired by the simple fall of an apple, a moment that sparked his curiosity about why objects fall downward. He realized that the same force pulling the apple to the ground might also extend to celestial bodies, keeping the Moon in orbit around Earth and Earth bound to the Sun. This insight laid the foundation for his groundbreaking work, Philosophiæ Naturalis Principia Mathematica, where he introduced the law of universal gravitation.

Newton's law described gravity as a force acting at a distance, proportional to the masses involved and inversely proportional to the square of the distance between them. It elegantly explained phenomena ranging from the tides to the orbits of planets. Yet, it left one crucial question unanswered: how does gravity act across empty space? For over two centuries, this puzzle remained, even as Newton's equations continued to guide astronomers in mapping the heavens and engineers in designing structures on Earth. Gravity was understood as a predictable force, but its true nature remained a mystery.

Enter Albert Einstein in the early 20th century, who transformed our understanding of gravity with his theory of General Relativity. Einstein envisioned gravity not as a force but as a curvature of spacetime caused by mass and energy. Massive objects like planets and stars bend the fabric of spacetime, creating the effects we perceive as gravitational attraction. This theory not only confirmed Newton's observations on a broader scale but also unveiled new phenomena, such as black holes and the bending of light around massive objects. Today, gravity continues to intrigue scientists, connecting us to the fundamental workings of the universe, from the fall of an apple to the dance of galaxies billions of light-years away.

# Why Can Bears Hibernate but Not Humans?

Bears hibernate because nature has equipped them with an extraordinary physiological toolkit that allows them to survive months of harsh winter with limited food. During hibernation, a bear's metabolism slows dramatically, reducing its energy needs by up to 75 percent. Its heart rate drops to as low as 8 beats per minute, and its body temperature decreases slightly—not as much as in smaller hibernating animals, but enough to conserve energy. Remarkably, bears can recycle their own waste products during this time, avoiding the muscle and bone loss humans would suffer in prolonged inactivity. This combination of adaptations allows bears to weather food scarcity while maintaining their health and strength.

Humans, on the other hand, are not biologically designed for hibernation. Our metabolism is not capable of the extreme slowdown seen in bears, and we cannot recycle proteins and other essential nutrients the way hibernators do. If a human were to attempt months of inactivity, the consequences would be dire. Muscle atrophy, bone density loss, and organ failure would occur without the body's regular replenishment of nutrients. Furthermore, our brain, which consumes a significant amount of our energy, lacks the mechanisms to enter the prolonged, low-energy state that allows hibernating animals to survive. Evolution simply did not shape us for such an extreme survival strategy because humans developed other ways—like agriculture, shelter, and clothing—to endure seasonal changes.

From a broader perspective, hibernation in animals like bears is a marvel of evolutionary engineering tailored to specific environmental challenges. Humans, while lacking the ability to hibernate, excel in adaptability and ingenuity, having found other ways to dominate nearly every ecosystem on Earth. Yet, the study of hibernation continues to intrigue scientists, who hope to one day apply its principles to fields like space travel and medicine. Imagine astronauts entering a hibernation-like state for long journeys to distant planets or patients recovering from trauma with reduced metabolic needs. While we may not hibernate like bears, understanding their biology might unlock profound possibilities for the future.

# How Do Leaves Change Color?

The changing colors of leaves in autumn is one of nature's most beautiful transitions, and it all starts with the chemistry inside the leaf. During spring and summer, leaves are packed with chlorophyll, the green pigment that helps plants capture sunlight for photosynthesis, turning light into food. As the days grow shorter and temperatures drop in autumn, trees prepare for winter by slowing down their food production. Chlorophyll breaks down and fades away, revealing other pigments that have been there all along but were hidden by the dominant green.

These pigments create the dazzling reds, oranges, and yellows we associate with fall. Carotenoids, which produce yellow and orange hues, are the same pigments that give carrots their color. Anthocyanins, responsible for the reds and purples, are produced by some trees in autumn and act as a kind of sunscreen for the leaf, protecting it from damage as it breaks down nutrients. The exact colors and intensity depend on the tree species and environmental factors like sunlight and temperature. Cool nights and sunny days often produce the most vibrant displays.

The annual transformation of leaves is more than a spectacle—it is an essential part of the tree's life cycle. By shedding their leaves, trees conserve water and energy during the harsh winter months. The fallen leaves enrich the soil as they decompose, contributing to the ecosystem's health. This process is a reminder of nature's intricate systems, where even the simplest changes, like the breakdown of chlorophyll, play a role in the larger rhythms of life. Autumn leaves are not just beautiful; they are evidence of nature's ability to adapt and thrive in changing conditions.

# What Causes Thunder And Lightning?

Thunder and lightning are dramatic displays of nature's raw energy, born from the intense conditions inside a thunderstorm. Lightning begins when powerful updrafts of warm air and downdrafts of cooler air within a storm cloud create friction, causing tiny ice particles and water droplets to collide. These collisions generate static electricity, separating the cloud into regions of positive and negative charges. When the difference between these charges becomes too great, the cloud releases a sudden discharge of electricity—a lightning bolt. This immense energy travels through the atmosphere, seeking balance by neutralizing the charges.

A lightning bolt heats the air around it to temperatures as high as 30,000 Kelvin—hotter than the surface of the Sun. This rapid heating causes the air to expand explosively, creating a shockwave. When the shockwave reaches your ears, you hear it as thunder. Because light travels faster than sound, you see the flash of lightning before you hear the accompanying rumble of thunder. The delay between the two can even help you estimate how far away the storm is; each three-second gap represents about one kilometer, or roughly five seconds per mile.

Thunder and lightning are more than just natural fireworks—they are key players in Earth's weather and energy systems. Lightning helps maintain the balance of nitrogen in the atmosphere by breaking apart molecules, creating compounds essential for plant growth. It also reminds us of the incredible power of nature, turning simple water droplets and air molecules into one of the most awe-inspiring forces on Earth. Every flash and roar is a testament to the dynamic processes shaping our planet, inviting us to marvel at the science behind the storm.

# How Do Magnets Work?

Magnets work thanks to the fundamental forces of nature, specifically the interplay between electricity and magnetism. At the heart of a magnet's power lies its atomic structure. In most materials, the tiny magnetic fields generated by the movement of electrons—those negatively charged particles orbiting atomic nuclei—tend to cancel each other out. However, in certain materials like iron, cobalt, and nickel, the electrons' magnetic fields align in the same direction. This alignment creates a collective magnetic field, turning the material into what we recognize as a magnet.

The invisible force of a magnet is the result of this alignment, producing a magnetic field that extends into the surrounding space. The field has two poles, north and south, where the magnetic force is strongest. Opposite poles attract, while like poles repel, a phenomenon that underpins much of magnetism's practical applications. This field can even influence other materials nearby, temporarily magnetizing them if their atoms can align with the field. It is why a paperclip sticks to a magnet and why Earth's own magnetic field protects us from harmful solar radiation.

Magnets are more than just tools for sticking notes to the fridge; they are critical to modern technology and understanding the world around us. They enable the function of electric motors, generators, and countless electronic devices. Magnetic resonance imaging (MRI) uses magnets to peer inside the human body, and even our credit cards rely on magnetic strips to store data. From daily conveniences to groundbreaking advancements in science and medicine, magnets demonstrate how nature's forces can be harnessed to improve our lives.

With its cratered surface and dusty plains, the Moon tells a story written in the language of impact and time. Unlike Earth, it lacks an atmosphere to shield it from meteoroids, which is why its surface is pockmarked with craters, some billions of years old. The dark, smooth regions known as maria are vast plains of solidified lava from ancient volcanic activity. These features, visible even with the naked eye, make the Moon a geological time capsule, preserving the history of our solar system in a way Earth's dynamic surface cannot.

The Moon has also inspired humanity to reach for the stars, both literally and figuratively. From early myths and calendars to the Apollo missions, it has been a source of wonder and discovery. Today, it continues to beckon scientists and explorers, serving as a stepping stone for deeper space exploration. Plans to establish lunar bases and extract resources could transform our understanding of space travel and survival. The Moon is more than a celestial neighbor - it is a testament to Earth's history, a partner in shaping our world, and a gateway to humanity's future among the stars.

# How Are Mountains Formed?

Mountains, those towering sentinels of the Earth, are formed through immense geological forces that reshape the planet's crust. Most mountains arise from the movement of tectonic plates, the massive slabs of rock that make up Earth's surface. When these plates collide, their edges crumple and fold, much like a rug pushed against a wall. This process, known as orogeny, can thrust rock layers upward over millions of years, creating majestic ranges like the Himalayas, which continue to rise as the Indian and Eurasian plates press against each other.

Not all mountains are created through collision. Some are formed by volcanic activity when molten rock, or magma, rises from beneath the Earth's crust and solidifies. This is how volcanic mountains like Mount Fuji in Japan or Mount Kilimanjaro in Africa are born. Others, called fault-block mountains, result from the stretching and cracking of the Earth's crust. In these cases, large blocks of rock are pushed up or tilted as the crust pulls apart, creating steep, dramatic landscapes like the Sierra Nevada range in the United States.

Mountains are more than just geological marvels; they are vital to life on Earth. They influence weather patterns by forcing air to rise, cool, and condense, which generates precipitation and feeds rivers. They provide habitats for diverse ecosystems and are a source of valuable resources, from water to minerals. Mountains also remind us of the immense forces shaping our planet, forces that operate on timescales far beyond human experience. Each peak and ridge is a testament to Earth's restless nature, sculpting landscapes that inspire awe and curiosity in equal measure.

# What is Photosynthesis?

Photosynthesis is one of the most extraordinary processes in nature, a biochemical symphony that powers life on Earth. At its core, photosynthesis is the way plants, algae, and some bacteria convert sunlight into energy. Using chlorophyll, a green pigment found in their cells, plants capture sunlight and combine it with water from the soil and carbon dioxide from the air. This reaction produces glucose, a sugar that serves as a vital energy source, and oxygen, a byproduct released into the atmosphere. It is a system so efficient that it transforms the raw energy of the Sun into the foundation of nearly every food chain on the planet.

The process happens in two main stages: the light-dependent reactions and the light-independent reactions, also known as the Calvin cycle. In the first stage, sunlight drives the splitting of water molecules, releasing oxygen and creating energy-storing molecules like ATP and NADPH. In the second stage, these molecules fuel the conversion of carbon dioxide into glucose. This delicate interplay of physics, chemistry, and biology occurs in chloroplasts, specialized structures within plant cells. It is an astonishing example of how living organisms harness the fundamental forces of nature to sustain themselves and their ecosystems.

Photosynthesis is not just a plant's way of making food - it is Earth's life support system. The oxygen you breathe and the energy you consume both trace back to this remarkable process. It regulates atmospheric carbon dioxide levels, playing a critical role in maintaining the planet's climate. Photosynthesis exemplifies the interconnectedness of life, illustrating how sunlight, air, and water combine to sustain the intricate web of organisms on Earth. It is a reminder of nature's elegance, where simple ingredients come together to create the complexity of life itself.

# How is a Rainbow Formed?

A rainbow is one of nature's most breathtaking optical phenomena, a vivid display of sunlight interacting with water droplets in the atmosphere. The process begins when sunlight encounters a raindrop. As the light enters the droplet, it slows down and bends, a phenomenon known as refraction. Inside the droplet, the light reflects off the back surface and exits, bending again as it leaves. This double refraction and reflection cause the white light to spread out into its constituent colors, creating the spectrum we see in a rainbow.

The colors of the rainbow are always the same, appearing in the order of red, orange, yellow, green, blue, indigo, and violet. These seven colors are often remembered by the acronym ROYGBIV. Each color corresponds to a specific wavelength of light, with red having the longest wavelength and violet the shortest. The arrangement of colors is not random—it results from the way light bends and separates as it passes through countless raindrops, with each drop acting as a tiny prism.

Rainbows are a beautiful reminder of the science hidden in everyday phenomena. They show us how the laws of physics and the properties of light combine to create something stunning. Beyond their aesthetic appeal, rainbows also symbolize the intricate interplay between the Sun, water, and our atmosphere. Whether glimpsed after a storm or during a gentle rain shower, a rainbow invites us to pause and marvel at the natural world's ability to transform simple interactions into extraordinary displays.

# What Are Rain Clouds?

Rain clouds, those dramatic harbingers of storms and life-giving water, are an extraordinary example of nature's ability to transform invisible moisture into visible wonders. They form when warm, moist air rises and cools, causing the water vapor in the air to condense into tiny droplets or ice crystals. These droplets cluster together around microscopic particles like dust or pollen, creating what we see as clouds. The specific types of clouds that produce rain are nimbostratus and cumulonimbus clouds, both known for their capacity to bring precipitation in various intensities.

Nimbostratus clouds are thick, gray layers that cover the sky like a heavy blanket. They often bring steady, widespread rain or snow, lasting for hours or even days. In contrast, cumulonimbus clouds are towering and dramatic, reaching high into the atmosphere. These clouds are associated with intense weather, such as thunderstorms, heavy rain, and even hail or tornadoes. Their vertical development makes them some of the most powerful and visually striking rain clouds.

Rain clouds are an integral part of the Earth's water cycle, redistributing moisture across the planet and sustaining ecosystems. They connect the oceans, land, and atmosphere in a continuous loop of evaporation, condensation, and precipitation. Without rain clouds, life as we know it would not exist—no rivers, no crops, no forests. Each raindrop falling from the sky is a testament to the intricate balance of forces that govern our planet, a reminder of the profound interconnection between the atmosphere and the life it nurtures.

# What Causes Seasons?

Seasons, those rhythmic changes in weather and daylight, are caused by the tilt of Earth's axis as it orbits the Sun. Our planet's axis is tilted at an angle of about 23.5 degrees, which means different parts of Earth receive varying amounts of sunlight throughout the year. This tilt, combined with Earth's constant revolution around the Sun, creates the annual cycle of seasons. When the Northern Hemisphere is tilted toward the Sun, it experiences summer, with longer days and more direct sunlight, while the Southern Hemisphere undergoes winter, with shorter days and less direct sunlight. Six months later, the situation reverses. It is crucial to understand that seasons are not caused by how close Earth is to the Sun. In fact, Earth is slightly closer to the Sun during the Northern Hemisphere's winter.

Earth experiences four distinct seasons: spring, summer, autumn (or fall), and winter. Each season lasts about three months and is marked by characteristic weather patterns and environmental changes. Spring is a time of renewal, with blooming flowers and rising temperatures. Summer brings the warmth and long days of peak sunlight. Autumn is the season of cooling temperatures and falling leaves, while winter brings the coldest weather, shorter days, and often snow in many regions. These transitions are driven by Earth's orbit and the tilt of its axis, creating a regular cycle that has shaped life on the planet.

Seasons are more than just a backdrop to our lives—they are deeply tied to the cycles of nature and human activity. They regulate the growth of crops, the migration of animals, and even cultural traditions around the globe. From the budding flowers of spring to the snowy landscapes of winter, the changing seasons remind us of the dynamic relationship between Earth and the Sun. It is a dance of celestial mechanics that shapes our world in profound and beautiful ways.

# How Shadows Are Formed?

Shadows are formed when an object blocks the path of light, creating a region of darkness where the light cannot reach. Light travels in straight lines, so when it encounters an opaque object—one that does not allow light to pass through—it casts a shadow on the surface behind it. The shape of the shadow matches the outline of the object, distorted only by the angle and distance of the light source. This phenomenon is a simple yet profound demonstration of how light interacts with matter.

The characteristics of a shadow depend on the size and position of the light source. A smaller or more distant light source creates a sharp-edged shadow, while a larger or closer light source produces softer edges, a phenomenon called the penumbra. Shadows also vary in size depending on the angle of the light. When the Sun is low in the sky, such as during sunrise or sunset, shadows stretch long and dramatic. When the Sun is overhead at noon, shadows are shorter and more compact. These variations show how geometry and light work together to create the shadows we see every day.

Shadows are more than just visual effects; they help us understand the physical properties of light and objects. Artists use shadows to create depth and perspective, while scientists study them to learn about celestial bodies and their orbits. For example, eclipses are essentially large-scale shadows cast by planets and moons. Shadows are also part of our daily lives, marking time with sundials and offering cool relief on a sunny day. They remind us that light, despite its speed and power, cannot illuminate everything - a balance that adds depth to both science and our perception of the world.

# What Causes Snow?

Snow is one of nature's most intricate phenomena, a crystalline transformation of water vapor into delicate ice flakes. It begins high in the atmosphere, where temperatures are freezing, and tiny water droplets or ice crystals come into contact with microscopic particles like dust. These particles serve as nuclei around which water vapor condenses and freezes. As more vapor accumulates, the frozen droplets grow into complex ice crystals, each with a unique six-sided structure dictated by the laws of physics. The process is a delicate interplay of temperature, humidity, and air currents.

When these crystals become heavy enough, they begin to fall toward the ground. As they descend, their journey is shaped by the atmospheric conditions they encounter. If the air remains below freezing all the way to the surface, the crystals reach the ground as snow. However, if they pass through a layer of warmer air, they may partially melt and refreeze, leading to sleet, or fall as rain instead. Snowflakes vary widely in size and shape depending on factors like temperature and moisture levels, with colder, drier conditions often producing smaller, more intricate flakes.

Snow is more than just frozen water—it is a key player in Earth's climate system. Snow cover reflects sunlight back into space, helping to regulate global temperatures. It also provides vital water resources when it melts in the spring, replenishing rivers and aquifers. Beyond its scientific significance, snow inspires awe and wonder with its beauty and transformative power. Each snowfall is a reminder of the complex interactions between Earth's atmosphere and the water cycle, showcasing how even the smallest particles can come together to create something magnificent.

# How Stars Are Formed?

Stars are born in vast clouds of gas and dust known as nebulae, the stellar nurseries of the cosmos. These regions are filled with hydrogen, helium, and trace elements, gently drifting until gravity begins to work its magic. Over time, regions within the nebula grow denser, pulling in more material under their own gravity. As this happens, the core of the forming star, or protostar, becomes increasingly hot and pressurized. When the temperature in the core reaches about 10 million Kelvin, nuclear fusion ignites, fusing hydrogen atoms into helium and releasing tremendous energy. At this moment, a star is born, shining brightly as it begins its main sequence of life.

The color of a star reveals its temperature, a cosmic thermometer painted across the sky. Blue stars, like Rigel in the constellation Orion, are the hottest, with surface temperatures exceeding 30,000 Kelvin. White stars, such as Sirius, are slightly cooler but still blaze intensely. Yellow stars, like our Sun, have surface temperatures around 6,000 Kelvin, radiating a comforting warmth. Cooler stars, like Betelgeuse, glow orange or red, with surface temperatures below 4,000 Kelvin. These colors are not just aesthetic—they provide astronomers with clues about a star's age, composition, and energy output.

The formation of stars is an ongoing process that shapes galaxies and drives the cycles of matter and energy in the universe. As stars age, they create heavier elements through fusion, eventually dispersing them into space during supernovae or as stellar winds. These materials seed future generations of stars, planets, and even life itself. The story of a star's birth, color, and evolution is a testament to the dynamic processes that govern the cosmos, reminding us that the light we see in the night sky is a snapshot of both the past and the ever-evolving universe.

# The Sun

The Sun, our closest star, is a colossal ball of glowing plasma at the center of the solar system, radiating the energy that sustains life on Earth. At its core, the Sun fuses hydrogen atoms into helium through nuclear fusion, a process that releases immense amounts of energy. This energy travels outward through the Sun's layers and eventually escapes as sunlight, a mix of light, heat, and other forms of electromagnetic radiation. The Sun's core reaches temperatures of about 15 million Kelvin, making it the powerhouse that drives the cycles of life on our planet.

The Sun's influence extends far beyond providing warmth and daylight. Its gravitational pull keeps the planets in orbit, orchestrating the intricate dance of the solar system. The energy it emits drives Earth's weather patterns, ocean currents, and the process of photosynthesis, which forms the foundation of most life. However, the Sun is not constant; its surface, the photosphere, is marked by dynamic activity like sunspots, solar flares, and coronal mass ejections. These phenomena release bursts of charged particles that can disrupt satellites and power grids, reminding us of the Sun's immense and sometimes unpredictable power.

As a middle-aged star, the Sun is about halfway through its 10-billion-year lifespan. Over time, it will evolve, expanding into a red giant and eventually shedding its outer layers to become a white dwarf, leaving behind a legacy of elements scattered across the cosmos. The Sun is not just a celestial body—it is the giver of life, the architect of the solar system, and a reminder of the intricate forces that govern the universe. Its brilliance connects us to the vastness of space, anchoring our existence to the rhythms of the cosmos.

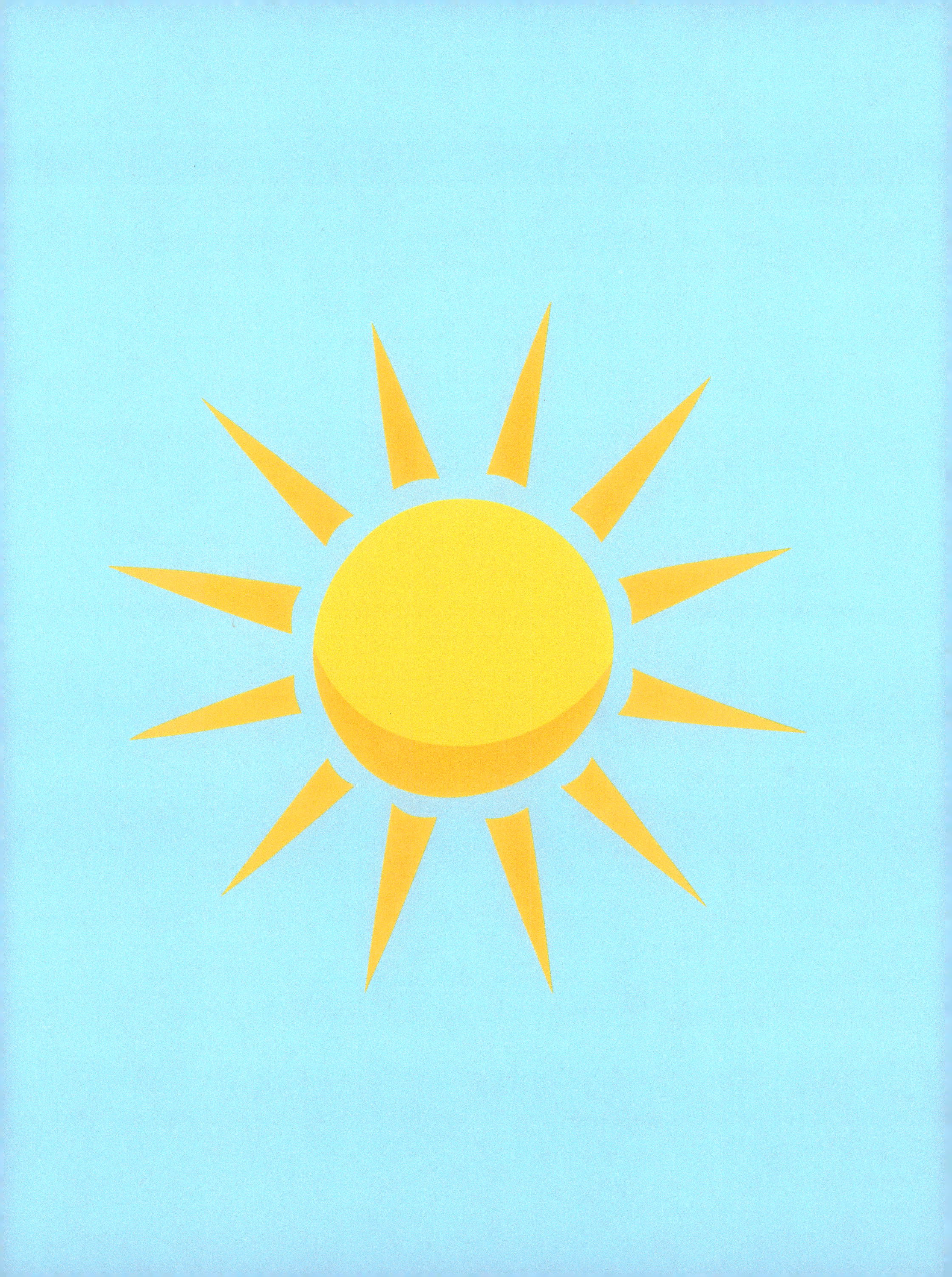

# What Causes Tornadoes?

Tornadoes, those awe-inspiring yet destructive spirals of wind, are born in the chaos of severe thunderstorms. They begin when warm, moist air from the ground collides with cool, dry air from above, creating an unstable atmosphere. This clash sets the stage for a powerful updraft, a rising column of air. If wind speeds at different altitudes blow in varying directions or at different speeds—a condition known as wind shear—it can cause the rising air to start spinning horizontally. The storm's updraft then tilts this rotating column upright, creating the swirling vortex we recognize as a tornado.

Tornado formation is most common in regions like the central United States, known as "Tornado Alley," where such atmospheric conditions frequently occur. These storms are categorized by the Enhanced Fujita (EF) scale, which measures their wind speeds and the extent of damage. A tornado's color depends on the light conditions and the debris it picks up. For example, a tornado in a dry field might appear brown, while one over water could look gray or white. Regardless of their appearance, all tornadoes are a stark reminder of the atmosphere's incredible power and complexity.

While tornadoes carve paths of destruction on Earth, stars of various colors illuminate the cosmos above. Blue stars are the hottest and youngest, while red stars are cooler and often older. Just as a tornado's intensity reveals the forces at play in the atmosphere, a star's color tells us about its temperature and life stage. These phenomena, one terrestrial and the other celestial, showcase nature's diversity and dynamic energy. From the spinning winds of a tornado to the blazing light of a star, both inspire awe and deepen our understanding of the forces that shape our world and beyond.

# How Are Tsunamis Caused?

Tsunamis, those colossal waves that can reshape coastlines, are caused by powerful disturbances beneath the ocean. Most commonly, they originate from undersea earthquakes, where tectonic plates shift and release immense amounts of energy. When one plate pushes upward during this seismic activity, it displaces a massive volume of water, setting the stage for a tsunami. Unlike regular ocean waves, which are driven by wind, tsunamis move through the entire depth of the ocean, traveling at speeds up to 800 kilometers per hour. As they approach shallow coastal areas, their speed decreases, but their height can grow dramatically, creating the towering walls of water that crash ashore with devastating force.

Other causes of tsunamis include volcanic eruptions, landslides, or even asteroid impacts. For instance, a collapsing volcanic island can trigger a massive wave by displacing water suddenly. While these events are rarer, their effects can be just as catastrophic. Tsunamis are not just a single wave but a series of them, sometimes arriving minutes or even hours apart. This makes them uniquely destructive, as the second or third wave can often be the largest, catching people off guard. They are a vivid reminder of the dynamic and interconnected nature of Earth's geological and oceanic systems.

As tsunamis ripple across the planet's surface, stars blaze across the cosmos, displaying their colors as a clue to their temperatures and life stages. Blue stars, the hottest and youngest, radiate immense energy, while red stars, cooler and often older, glow with a steadier light. Like tsunamis, stars represent the immense power of natural forces—one sculpting our planet's surface, the other shaping the universe's structure. Whether it is the roaring energy of Earth's oceans or the serene light of a distant star, both phenomena showcase the beauty and power of nature's grand designs.

# How Do Volcanoes Form?

Volcanoes form when intense heat and pressure from inside the Earth create openings in its surface, allowing molten rock to escape. Deep beneath the crust lies the mantle, a layer of semi-liquid rock called magma. This magma is driven upward by immense heat from the Earth's core and the movement of tectonic plates—the massive, shifting slabs of rock that make up Earth's surface. When these plates pull apart, collide, or slide over one another, they create pathways for magma to rise, often forming volcanoes along their boundaries.

There are different ways volcanoes form, depending on the tectonic activity. At divergent boundaries, where plates move apart, magma rises to fill the gaps, creating volcanoes like those along the mid-ocean ridges. At convergent boundaries, where one plate slides beneath another, the sinking plate melts as it gets deeper, producing magma that forms volcanoes like the ones in the Pacific Ring of Fire. Some volcanoes, like those in Hawaii, form over hotspots—places where plumes of hot magma rise directly from deep within the mantle, independent of tectonic boundaries.

Volcanoes are not just geological curiosities; they are essential to shaping the planet. Their eruptions build mountains, create new islands, and enrich soils with nutrients that support life. However, they also remind us of the Earth's immense power, capable of both destruction and renewal. Each volcano tells a story of the forces at work beneath our feet, connecting us to the dynamic processes that have been shaping our planet for billions of years.

# What is the Water Cycle?

The water cycle, or hydrologic cycle, is Earth's masterful system for moving water through its environment, sustaining life and shaping landscapes. It begins with evaporation, where the Sun's energy heats water from oceans, lakes, and rivers, turning it into vapor. This invisible vapor rises into the atmosphere, joined by water released from plants through transpiration, a process where vegetation "breathes" moisture into the air. Together, these steps transport water from Earth's surface to the skies, forming the starting point of this continuous cycle.

As the vapor ascends, cooler temperatures in the atmosphere cause it to condense into tiny droplets, forming clouds—a phase called condensation. These clouds are carried by wind until the droplets grow large and heavy enough to fall as precipitation. Depending on the conditions, this precipitation can take the form of rain, snow, sleet, or hail. This step returns water to Earth's surface, replenishing oceans, rivers, lakes, and groundwater reservoirs. Precipitation also feeds plants and animals, sustaining ecosystems in the process.

Once on the ground, water continues its journey in various forms. Some of it flows across the surface as runoff, collecting in streams and rivers that eventually return to the oceans. Other portions seep into the ground, recharging aquifers in a process called infiltration. This groundwater may remain stored for years or emerge later as springs, completing its return to surface water sources. The water cycle is not just a process but a lifeline, connecting the atmosphere, oceans, and land in a seamless loop. It showcases Earth's ability to recycle its most precious resource, ensuring a steady supply of water to support life across the planet.

# What Causes Ocean Waves?

Ocean waves, those mesmerizing undulations of water, are born from the dynamic interaction between wind and the sea. Most waves are created when wind blows across the surface of the ocean, transferring energy to the water through friction. As the wind pushes against the water, small ripples form and grow larger as the wind continues to exert force. The stronger and more sustained the wind, and the longer the distance it travels uninterrupted across the water—called the "fetch"—the larger the waves become. This is why storms over open seas can generate enormous swells.

The energy within a wave moves forward, but the water itself does not travel with the wave. Instead, the water particles move in circular orbits, rising and falling as the wave passes through. This orbital motion decreases with depth, making waves mostly a surface phenomenon. When waves approach the shore, they encounter shallower water, which disrupts their circular motion and causes the waves to slow down, grow taller, and eventually break. This breaking action creates the iconic cresting waves that crash onto the beach, releasing their stored energy.

Ocean waves are not just a spectacle but a vital force in shaping our planet. They help erode coastlines, transport sediments, and distribute energy across the oceans. Beyond their geological impact, waves play a role in climate regulation by driving ocean currents that circulate heat and nutrients. They also inspire human ingenuity, from surfing to harnessing wave energy for sustainable power. Each wave is a reminder of the interconnected forces of nature, driven by the interplay of wind, water, and the ever-present energy of our planet's systems.

# Glossary

*Aerodynamics* is the study of how air moves around things, like how birds fly, airplanes soar, and kites dance in the sky!

The *atmosphere* is the invisible blanket of air around Earth that keeps us safe, gives us air to breathe, and makes the sky blue!

Adenosine Triphosphate, or *ATP*, is like a tiny battery in your body that gives your cells the energy to do their jobs!

*Bioluminescence* is when living things like fireflies and glowing jellyfish make their own light, like a natural flashlight!

*Chlorophyll* is the green stuff in leaves that helps plants soak up sunlight to make their food!

*Coalescing* means coming together to make one big thing, like tiny drops of water joining to form a puddle!

*Colossal* means something super big, like a giant dinosaur or a really tall mountain!

*Coronal* means something that comes from the outer part of the Sun, like the bright, glowing crown of light you see during a solar eclipse!

*Figuratively* means saying something in a fun or imaginative way, like "I'm so hungry I could eat a horse," but you do not really mean it!

*Hibernate* means going into a deep sleep for a long time, like bears do in winter to stay cozy and save energy!

*Kelvin* is a special way scientists measure temperature, starting from the coldest cold ever, where everything is completely frozen, called absolute zero!

*Metabolism* is the collection of chemical processes that occur within a living organism to maintain life, involving the breakdown of nutrients for energy and the synthesis of necessary compounds.

Nicotinamide adenine dinucleotide phosphate, or *NADP*, is like a little helper in plants that carries energy to make food during photosynthesis!

A *nebula* is a big, colorful cloud of gas and dust in space where stars are born or where old stars leave behind their glowing leftovers!

*Neutralizing* means making something calm or balanced, like when two kids on a seesaw even each other out so it stays level!

A *phenomenon* is something amazing or interesting that happens, like a rainbow in the sky or leaves changing color!

*Physics* is the science of how everything moves, works, and interacts, from rolling balls to shining stars!

*Profound* means something very deep or important, like a big idea that makes you think about life in a whole new way!

A *spectacle* is something amazing or exciting to see, like a rainbow in the sky or fireworks lighting up the night!

www.ingramcontent.com/pod-product-compliance
Lightning Source LLC
Chambersburg PA
CBHW052042190326
41519CB00003BA/254